I0480039

VOLUME 22

THE ERROR OF THE GREAT SCIENTISTS

EXTENDING THE THEORY OF BIG BANG

FIRST EDITION

Carlos L. Partidas

N° Depósito Legal   MI2019000454
ISBN: 978 1670 9187 10

9 781670 918710

REGISTRATION OF INTELLECTUAL PROPERTY SAPI: N° 8074
OF THE COMPENDIUM THE CHEMISTRY OF DISEASES
BOLIVARIAN REPUBLIC OF VENEZUELA, 07/05/2010

DEDICATORY

To the memory of the great scientists, who have helped to clear the intricate path of science, so that humanity may pass freely along the path of knowledge.

CONTENTS

RECOGNITION

To the energetic force of the Almatrinos, who made it possible from before time zero, the creation of our immense Universe

1

THE FINAL HISTORY OF PHILOSOPHY

During one of Google's Zeitgeist Conferences in 2011, Stephen Hawking stated that "philosophy was dead". Hawking said, "Philosophers have not kept pace with advances in science, while scientists have become the bearers of the flame of discovery. And Hawking added that "philosophical doubts can be clarified by science, and in particular by new scientific theories, which show us a different image of the universe ". But this is undoubtedly a great success of Stephen Hawking, because science only takes as certain an explanation, provided that this hypothesis can be tested experimentally. Therefore, the idea that is glimpsed based on a certain theory, will pass through a series of tests, which will be subject to errors and successes. However, some scientists who try to elaborate an accurate explanation with the support of experimental data, rely only on matrices that arise from mathematics, and forget to focus their gaze on the real phenomenon. And they believe only in the result that the calculations predict. But this is a mistake that the great scientists have made, who also because of their prestige, have managed to get other thinkers to follow them blindly, without the need to establish their own criteria or without objections. And we can say as Stephen Hawking, that in reality philosophy is dead, because it is science that is unveiling the great mysteries of Nature. But this discipline is to be understood that it was useful, because it was the one that forced the human being to think about how to give an explanation to its origin; to the enigmas and to the real principle of its cosmic world.

Or perhaps, because in the old days the necessary instruments for carrying out experimental methods had not been developed, so that explanations of phenomena arose only from the capacity of thought, which made the formation of philosophical thinkers burst forth, as did the emergence of great scientists. Or perhaps that is why it has corresponded to some of us to observe more closely the reconstruction of phenomena, based on data or on the tools given to us by new and modern instruments, to follow the indelible traces left by the evolution of the Universe.

Or we can say that when experimental methods did not yet exist, the ideas of the philosophers acted with great impetus, because they were the ones who could explain, each in their own way, the great mysteries of the origin of the Universe. For example, until the seventeenth century, it was considered that the tendency of a body to fall to the earth was a property inherent to all bodies. Therefore, the phenomenon was thus clarified, and therefore no other explanation was required. Until William Stukeley in his book "Memoir of Sr Isaac Newton's Life" published in 1752, describes that he met the great scientist drinking tea in a garden under apple trees, and that when Newton saw an apple fall, he commented to Stukeley that that scenario was the same as when he described the idea of grav-itation. And Stukeley wrote: "An apple fell on him when he rested meditating"... Although we know that gravitational force is a property of bodies that has to be measured, but can-not be predicted by a mathematical function.

And among that series of philosophical confusions, it would be that the enormous number of churches arose, when some philosophers coincided, that in the initial point of all this, there must have been the action of a creator. But the only thing that until now has remained hidden behind any experiment, has been the image of that creator. And this secret was also a great idea, because it is supposed that it is not going to be possible to demonstrate something that does not really exist. Therefore, it is the only thing that has not left a trace of its evidence, and in this way, the idea of the existence of a creator can be kept alive.

But there also arises the idea of atheism among the great sci-entists, when they observe that there is no evidence of the existence of a supreme being. And that search also reaches out to the great religious, when they try to prove that exist-ence in some way. Or logic can carry scientists as well as great philosophers and religious people on a raft in a troubled sea. And each one will decide if he prefers to become a philoso-pher when he or she tries to look for something in reality that he doesn't get. Or when he gets it, the storm will finally reach its calm, and inevitably, one idea will erase the other. However, as Stephen Hawking said, one of them, philosophy, is dead, because it can no longer investigate ideas on the raft with the rough sea, because science has calmed it down with experi-mental proof of every phenomenon, and the analytical mind of the human being is now rowing towards a safe harbor. While the religious, remains confident in his expectation.

But as for scientists and the formation of the Universe, perhaps this calm came when Edwin Powell Hubble observed that galaxies are moving away from Earth. And this assumes a reality, as it is, that the Universe is in a stage of growth, when even being in full tempest, all the rafters believed that the Universe was static, and that the center of the Universe was the Earth.

But this new idea of Hubble, or the reality of an expanding Universe, indicated that there should be a point of origin, from which it began to form into the Universe. And it was this idea that was proposed precisely by a reverend of the Catholic Church of Belgian origin, Georges Henri Joseph Édouard Lemaitre. Because this fact of a Universe in growth, suited him perfectly well to the search of the church; since it was supposed, that someone had to be behind that growth, to stoke that point and that the Universe was formed. Therefore, some non-religious scientists and cosmologists suspected that the church was meddling in those phenomena that could only be explained by science.

But then, this idea was bouncing and at the same time filling in some way the raft of events, until the idea of Georges Lemaitre, was accepted by most scientists and cosmologists, and all agreed to call it, the Big Bang Theory. Because this theory fits perfectly into the explanation of how matter originated from energy. But again, that this theory moved away from the idea of the religious, who still do not get their creator by means of this proposition. So, again, ideas are raised that many religious no longer share with scientists and cosmologists, because the Big Bang theory itself does not prove the existence of a creator. And it seems that God has to appear on the scene as an obligatory fact, or that he manages to please the immense number of religions, in spite of the fact that the human race is unique, and that is sailing on the same raft. Therefore, there should be no setbacks, to see who is really right. Because in the end, whatever the answer, the human race will remain the same human race, without the need for a preference for one or the other.

However, there are still those doubts which the Big Bang theory has not been able to clarify; and scientists also make mistakes when they try to clarify these doubts. For example, if we say that the point had a high density of matter, because philosophically it is supposed that all that matter was concentrated in a single point. But in addition, the Big Bang supposes that the energy arose because that point was extremely hot. So the big doubt is: where did that energy come from that heated that point? Or how was it that matter managed to integrate up to a high density at that point?

But it is here, where the errors of the great scientists arise, because experimentally it can be demonstrated, that when particles move at a great speed, they themselves create mass. And this phenomenon was what Albert Einstein's theory of relativity proved experimentally. But Albert Einstein put a stop to himself, because he focused only or to look at the phenomenon of the creation of mass, as in a concept that seems to be instead of a rather philosophical scientist, because Einstein dedicated himself to analyze this phenomenon, only from the mathematical point of view, and not in the scientific reality, that a particle in movement creates its mass.

Or let's say, Einstein only saw with his equation, the moment at which that motion of a particle is less than the speed with which light moves. But Einstein did not consider the mass that forms when the particle moves faster than light. Perhaps, because in Albert Einstein's reasoning, the idea was still settled, that the Universe was static, and only the particles that moved faster were those of light, which traveled in the form of beams called photons. And this was so for Einstein, because mathematics indicated to Einstein that if particles moved faster than light, then the mass created would be imaginary, which was one of Albert Einstein's great errors.

But Albert Einstein succeeds in getting this error to be taken away by Wolfgang Pauli, Georges Lemaitre, Peter Higgs and Stephen Hawking, just to name these four, as the most famous scientists, because with their ideas, they changed the old way of thinking, or the concept that humanity had about the origin of the Universe. And the boson of Peter Higgs, continues being the hope for the religious, who patiently, will continue waiting for their creator in the raft.

But it is by observing the logic of the phenomenon that the mystery can be solved, but not blindly with the concept embodied only in mathematics. And it was from Einstein's equation, that we have deduced an equation that explains in a clearer or more evident way, how the Universe was formed from nothing. Because all that was needed was a very small particle that began to move with a spiral twist form, and from this one, others were formed, that still could not be manifested as energy, because that space was too small. And those particles that still exist, we have had to call almatrinos, because they have the physical property, if we can call it that, of not having rest mass. And with the new concept of virtual numbers, we can say that these particles are so small that they won't be able to be detected. But this answers one of the doubts that cannot be explained by the Big Bang theory, such as the existence of 74% of the undetectable energy of the Universe, and 22% of the mass that cannot be detected either. And that is why it is called energy and dark mass respectively. And this equation, is what we represent of the form:

$$\mathrm{U} = m_0 C^3 / E$$

Which is more logical, because when the energy E was very small, the tangential velocity U of the particle became infinite, and the mass m was formed from the resting mass m_0. And as is reasonable, it is from the virtual numbers that we can say, that in this equation, m_0 was not zero, but was something too small, or less than zero. And it is from here, or at this precise moment, that we begin to refer to quantities or values that mathematically can be too small or undetectable.

For example, a quantum of electric charge is so small that we are not going to be able to detect it by means of the disintegration of the electricity, no matter how meticulously this fractionation is made. Or that we still do not realize that the air that enters through our nose, is formed by molecules, and that we can only perceive sometimes without objecting, that this substance is air. But let us suppose, for example, that in a bulb of 110 volts and 100 watts, elementary charges enter through the filament 6×10^{18} per second. So it's a real problem to imagine a world as small as the Almatrinos. But it is with the concept of virtual numbers that we can now move in such a large space, as the range from the minus infinite to the plus infinite $(-\infty, +\infty)$.

In such a way, that scientists have done their best to clarify the mysteries behind the creation of the Universe, but despite the billions of years that have passed since then, scientists discovered that there is a logical sequence, or that what remains is a trace after each event. In such a way, that any event occurred in that lapse, will leave an image like a trace, that can be used to develop a mathematical model, with which that trace can be left definitively printed, to be able to explain, how it was that event evolved.

And for this modeling, the invention of mathematics has been very useful, because it is the mathematical model, which allows us to capture or engrave on paper, or as a print or seal, the form of how events could have happened, so that we can then sit down to contemplate, analyze or imagine in retrospect, how that event happened, so that we can then project it towards any moment; or even towards a moment that is ahead in time.

But with the concept of time, as with mathematics, this is only an auxiliary element in science, since we cannot say that there are mathematics or time. Because we will not be able to weigh them or grasp them in a physical way. For example, we will not be able to hold in our hands a mathematical function, nor two seconds of time to know how they are or how much they weigh. But mathematics automatically builds all the combinations of numbers that we can assume, or those that we cannot imagine, because we only have to discover these intricate combinations. And in that incessant search among mathematics, we can be taken by shortcuts, or we will not be able to explain something that does not really exist.

And in terms of time, every event that has already happened can no longer happen in the same way. So it will be impossible to go back to a configuration of the past, because it will not happen again or be in the same form. And this was another mistake that Stephen Hawking made, when he stated that we should be careful when we travel back in time, because when we meet our origin, we could definitely die. So that would be like energetic suicide, which is totally illogical or impossible.

But the real thing is that we advance at a rate from an origin that we can already imagine, and towards an unknown end, although we know that this end is in the plus infinite ($+\infty$). But we say unknown, because we will not know how these events are going to happen. For example, humanity is destroying the Earth, but that is not the fault of the Universe. Although what we are not going to be able to predict exactly, is how, or what those consequences will be, in terms of the equilibrium of the solar system.

And everything that happens in that range will be unforeseen, because we will only be able to adapt to an impulse or rhythm imposed by the development or growth of the Universe, which is going to a compass that we are not going to be able to stop. Because in this event, only two forms are produced that can be felt and measured: one is energy and the other variable is distance. Since in the event, the Universe moves away in each fraction from its point of origin; but at the same time, it feeds itself with the energy it creates itself; and it only requires that the Universe be in constant movement.

And as for the human being, they only lives with a zero referential speed with respect to a body that is moving at the same rhythm as the Universe. Since the human being only rides on the body of the Earth, which is heading towards an unknown course. But we know that it is in the plus infinite. And by standing or moving with zero speed with respect to the Earth, this offers us opportunities to do something at this moment, and at this point in the Universe, from where we perceive that the Universe is still.

Or, for example, it is what allows the human being to be able to measure a certain lapse, which he calls time. And with this concept of time the idea of a static Universe takes root even more, because one lives with an illusion in a cyclical way. Or the human being lives trapped within his own created ideas. That is to say, enclosed only in the idea of time and mathematics. And he believes, for example, that events repeat themselves. So we can always celebrate Christmas, but that Christmas is not the same, because Christmas happened only once. Or it's impossible to live again on the same Monday, or the same Saturday, because Monday and Saturday don't exist, but printed in a cyclical way, on a cardboard called a calendar.

But perhaps, that idea came up, because the rotation of the Earth gives us the illusion that there is day and night, when in reality, what we are rotating is fixed at the same point that passes through another point where there is only light, because there is no shadow there, and then by another, where there is no light because what is there is only shadow.

2

THE GEOCENTRIC IDEA

Perhaps that the fixed location of a point on Earth was a key deduction, or one that fit the logic of the imagination, for the ancient Greeks to think that the Sun revolved around the Earth. Because if we tried to see the phenomenon during the night, we would be certain that the Earth is the one that revolves around the Sun, but it is impossible for us to see the Sun when we are passing through the zone of darkness. But if we could run in the same direction and with the same speed at which the Earth revolves around the Sun, then we would live eternally in the point of illumination. Or that to an observer standing on the surface of the Earth, a geostationary satellite, it would be perceived as if the satellite were situated at a stationary point in the illuminated sky. Or we would no longer be living at zero speed with respect to the Earth, but moving at the same speed as the Earth around the Sun. In other words, it actually seems that we are like a floating point on the surface of the Earth. And this is a geostationary displacement technique, which is used precisely in satellites, and it gives us the feeling that satellites are fixed in a point with respect to the Sun; or in which, the satellite remains static or standing on that point. So, for someone who is riding on the satellite, there will be no day or night. But this is achieved by making the satellite move in the same direction and with the same speed as the Earth with respect to the Sun. Or if you want, get on and try to climb an escalator, where the steps go up. And if you try to go down at the same speed as you go up the step, you will notice that it looks like you are standing on the same step, but you are not going up because you are floating. Or the other example, is when you are jogging on a conveyor belt; and in fact you are jogging but without moving from the site, because what moves is the conveyor belt.

And it is in this way that the Sun was thought to revolve around the Earth. An idea that comes from ancient Greek thinkers, or what is also called geocentrism, or rather geo synchronization. But it was this confusion that led astronomer Claudius Ptolemy in the second century to formulate a description of the conclusions of Greek astronomy, which is known as the Ptolemy hypothesis, or geocentric hypothesis. But it was the error of that reasoning that kept this idea alive for a long time. And because of this error, it was assumed that the Earth was fixed in the center of the Universe, while the Sun, Moon, and stars all moved around the Earth. And it was an idea accepted for almost a thousand five hundred years, which was enough to influence not only the way of interpreting science, but also astronomy and philosophy. But in the end this theory proved to be very complex, but in addition, it could not adapt to an ever-increasing number of observations by other thinkers. And this, without a doubt, was one of the errors that lasted the longest with the human race, and was committed by the Greek astronomer Claudio Ptolemy.

However, in the 16th century, Copernicus overturned the geocentric idea and suggested that a simpler description of celestial movements could be made, assuming that the Sun was fixed in the center of the Universe. And with this new theory of Copernicus, the Earth was only a planet revolving around the Sun, while the other planets had revolving movements similar to those of the Earth. And it was these controversies between the two theories that forced astronomers to investigate more closely the new idea of Copernicus' heliocentrism and Ptolemy's geocentrism. Such would be the case of Tycho Brahe, who would be the last great astronomer to carry out his research on heliocentrism, but the mistake is that Brahe did not have the help of a telescope.

Until in 1609, Galileo Galilei used a telescope built by him; and with that telescope, Galileo discovered the moons of Jupiter and the phases of Venus. Therefore, it was Galileo and not Brahe, who became the defender of Copernicus' ideas. Until about twenty years later, a Brahe assistant named Johannes Kepler, found some important evidence from the data of Tycho Brahe, about the movement of the stars. This made Johannes Kepler establish his three laws that considered the movement of the planets around the Sun. Or we can conclude that it was Copernicus who got his ideas from Claudius Ptolemy's error. But Nicolas Copernicus' mistake, like Brahe's, is that they didn't have a telescope to look and explore outer space, using a telescope to mark a fixed or reference point in space.

But this was also another mistake, because the erroneous idea that the Universe was static was long established in the minds of scientists. It even made Albert Einstein, who proposed in the law of relativity, to introduce a cosmological constant in order to explain why the Universe was static, make the same mistake. But Einstein retracted this idea in 1931, once Edwin Hubble observed the red shift of galaxies, which confirmed that the universe was not really static. And in 1930, Eddington demonstrated that the static Universe of relativity with a cosmological constant had no logic.

So this new constant was not justified, but it was proposed by Einstein, in order to obtain a result, which at that time was thought necessary. And when the evidence of Hubble's expansion of the Universe was presented, it is said that Einstein went so far as to declare that the introduction of such a constant was the "worst mistake of his life". And it was first written by physicist George Gamow in an article published in September 1956 in the journal Scientific American that Einstein's cosmological constant was a "blunder". But this publication was a year after the death of Einstein, who left the Earth, and we don't know where, in April 1955.

But like the Greeks, it was all part of a philosophy, until the experimental method proposed by Francis Bacon appeared. That is to say, it was philosophy that came out of the scene, when Galileo Galilei appeared with his famous telescope. And perhaps that is why Albert Einstein qualified Galileo as the father of experimental physics, because by being able to see towards an external point from the Earth, Galileo was able to experimentally prove that the Earth does indeed revolve around the Sun. And then William Herschel would emerge with a more powerful telescope than Galileo's. So, with this telescope, Herschel was able to see and explore a world farther away than we had previously imagined. Or even Herschel claimed that the Sun is actually an immense planet on which life exists, because he could see between the solar storms as they opened like curtains. But maybe that's where Albert Einstein lives with his energetic body.

But between the world of philosophy and science, is that we have moved with theories and experiments, to explain the mysteries of the Universe. Mysteries that, when discovered, turn out to be understandable as well as simple. But perhaps, that complexity is presented, when we try to capture the phenomenon or said mystery by means of a mathematical model. Because it is the same as languages; since through these, we still do not find how to express our feelings exactly, and we will have to use the gesture to help us express what we really feel. But we will not be able to write the gesture, to give or express an emotional content to the words written in our language.

And in the same way that for science, mathematical language is still full of imperfections, which does not allow us to explain a large number of phenomena, if we were based solely on mathematical language, without making a gesture towards the phenomenon. But supported by the imperfect language of mathematics, it is what has caused the great scientists to be guided by a series of errors, that perhaps someone, through the combination of philosophy and logic, that is, with thought and discernment, can explain cosmic phenomena in another way. Even without the need for mathematics. But in this way, a greater number of followers is also captured, when these followers do not have their own criteria. As it could be the case of the great quantity of religions that exist.

But again scientists fall into the error of thinking, that if something cannot be brought to a mathematical model, then it is because the phenomenon does not exist. Or without reasoning in the evidence of the phenomenon. But it is what forces us to introduce into the phenomenon, other terms, such as that the mass is imaginary. While some principles, such as Wolfgang Pauli's Exclusion Principle, are based on a logical interpretation, which would be more complicated to interpret, if we could take it to explain it through a purely mathematical language or model. However, we all accept Pauli's Principle of Exclusion through logic.

And when the human being thinks something to try to solve the problem of a certain phenomenon, science is the one that obliges they, so that through logic and experimental proof, thought has validity in deduction; or so that this idea can be captured by means of a mathematical function, and with which, a solution can be sought or a new law or a principle can be proposed that describes the phenomenon.

For example, the simplest mathematical function is $y=mx+b$. In such a way that any mathematician can deduce that this function corresponds to a straight line. Whereas a physicist would say that the phenomenon can be explained by a straight line. Because that is the dependence or relationship between the variable "y" and the variable "x". Since "m" is the slope of the line; and "b" is the point through which the line passes on the "y" axis. That is, we can draw the function on paper. And if "b" passes through the origin, then $b=0$ and the equation becomes more simply $y=mx$. And with this we will not be able to change the phenomenon, but only to explain it. And in order to explain more complex phenomena, the phenomena need to be represented by other more confusing functions.

And so, that each phenomenon will have its degree of confusion, until with the help of philosophy, we can solve the mystery of a phenomenon, which we don't value, when we can't explain it by means of a mathematical model. But the phenomenon will continue to exist. And it is in this concept that philosophers and religious people base themselves, who say that the fact of not being able to demonstrate the existence of God, does not mean in a categorical way that God does not exist. Only that it will still be invisible; or we won't be able to see it, because they philosophically say that God really is in everything that exists, so we can see to Him everywhere. And the religious says: you can't see Him, but there it is...

But also seen from the point of view of scientists, this paraphernalia between the philosophical and the scientific, is for example in explaining the movement of a single electron around a nucleus, such as the hydrogen atom. And it was Erwin Rudolf Josef Alexander Schrödinger, who wanted to take this simple phenomenon to a mathematical model. But this function is so complex that in the end, the purpose or idea of the mathematical function is not understood either. But the example is pathetic, because Schrödinger himself would not understand its mathematical function, because the only one who could understand it was Max Born, who could deduce that this function expressed the probability of finding the only electron in a given place and moment around the only hydrogen nucleus. So Max Born was awarded a Nobel Prize, which could have been for Schrödinger. But Schrödinger found it impossible or difficult to take his model to the helium atom, i.e. two electrons rotating around a nucleus with two protons. And that was undoubtedly Schrödinger's big mistake.

But perhaps the other one we can mention here is the case of the young Venezuelan mathematician Ramses Cornieles, who solved the problem of division by zero. But it was something that maybe Ramses did not understand either. However, he allowed to me to deduce what the Universe was like before time zero. But these perhaps are only some of the mistakes that the great scientists have made, because they are only going along the mathematical track, but they don't worry about looking for a solution, observing directly in the logic of nature and the reason why the phenomenon occurs that way. And what cannot be explained through mathematics is then given the title of "...mystery of science...", or all doctors who do not find the origin of a disease immediately attribute the cause of all that guilt to stress.

From there, the most outstanding ones arise; that is, scientists who have the capacity to analyze a phenomenon, but they cannot only leave it in their mind, but they have to take it to a mathematical model, for others to evaluate and value it; or for others to recognize or reject the idea. Therefore, in order to be able to graphically print the solution, they have to use mathematical language. It is like composing a melody, but we only know how to play it with a musical instrument. So it was necessary to learn how to write music on a staff, so that others can modify the melody, and play it, even if it is in a form similar to the original.

And in the same way, the scientist has to use mathematics to give coherence to his theory, or to be able to demonstrate that his thought has a solidly logical basis, or a valid sense. And if the proof can be repeated without errors, then probably philosophy perishes at that moment, while the theory comes to life, and will become or form part of a Law, and then if it has no objection, it will be a principle. Because laws can indeed be violated, but principles are inviolable. For example, the principle of fire is to burn, but fire cannot violate its principle of burning, and it does not matter if what is burned is a child or a forest with its beautiful fauna.

But it is up to other scientists to design complex experiments in order to check the validity of the theories of theoretical scientists, and they are called practical scientists. But an experiment can be very simple: for example, let two heavy balls roll over an inclined ramp, and if you are only attentive to the sound, you will notice that it is faster as the ball advances. Therefore, we deduce that another kind of force is acting on the spheres, which makes the speed of the spheres increase all the time. And we will call this force the force of acceleration of gravity, because if we test it hundreds of times, we will obtain the same result. And we will say that this invisible force is what causes the effect to be constant, and it is better to call it the Principle of Gravity. But this is the test that Galileo did.

And Galileo was also the first to try to know at what speed the light moves. Although his thought was certainly philosophical, and the only thing he had at hand to reference the speed with which light moves, was sound. In such a way that Galileo skillfully looked for an instrument, that is, a system that would allow him to see the light, but at the same time be able to hear the sound. And Galileo took the example of the cannon. But it was Galileo Galilei, the scientist who had to go as a rafter on that raft in the midst of the storm, because Galileo had to suffer the aggressions of the Inquisition imposed by the Catholic Church, which tried to warn against everything that interfered with their beliefs. Perhaps, because they hierarchy headed by the pope, understood that any argument made against the non-existence of God, could weaken its power. But apparently, the church had no choice but to accept those arguments that were unobjectionable, because science could prove them, provided that such theory sought in some way, involve or please them in reason, because they would continue in the search for evidence, which would give scientific support to their beliefs. And perhaps this was the case, and the great mistake made by the theoretical physicist of British origin, named Peter Higgs.

3

SCIENCE CLEARS ITS WAY

Because one of the most recent errors is precisely that of Peter Higgs; because he considered, that the discovery of the boson which he has theorized, should be called the particle of God, because his boson is supposed to have the integer value zero. Therefore, it is sensible to think that it was the boson that initiated the creation of the Universe; that is, that the Universe began to form at time zero with the boson zero. But here is the error of Peter Higgs, because if he was a supreme being who created that particle with a very high energy and an immense density, of course that creative energy could not have been a boson, because if it had been a boson, from that boson, would have formed a single particle. And it would be impossible to deduce, that after forming a boson can be divided, to generate from this the fermions. And if it had been so as Peter Higgs supposes, someone must have created this particle, but someone had to be the creator of that someone.

But it is also logical to suppose that this energy came out of nowhere, which is why Peter Higgs' theoretical model can't really explain to us how the Universe was formed. And it is a reality, that if the particles arose from nothing, then those particles were formed from an origin, where there was neither energy nor mass. In such a way, that the particles that gave origin to the Universe will not be able to be discovered by means of a detector or capturer of signals. And it is not because they are hidden behind a mystery, but technically, it is impossible to be able to detect these particles physically, because detectors cannot be built so that they "see" for us such small particles, by the only fact that these particles will be invisible to any detector that you want to build to detect them. And they would not be detectable for several logical reasons: for example, the design of the detector system would have to have particles of smaller size, or with a surface that is sufficient for the particles to rest and bounce. But that design, escapes any method or technical ability of the experiment.

And another example to compare, is that when we see the disk of the full Moon, it is because the electromagnetic waves that come out of the Sun in the form of light, bounce against the surface of the Moon, and in the rebound, the rays that come from the Sun, are reflected towards our sight. And the rough surface of the Moon causes the rays of light of the Sun to deviate, or to bounce separately; that is to say with a small offset in time and with a difference of intensity, thanks to the roughness of the lunar surface. In such a way, that because of this deface and different intensities, we can see places of lesser and greater luminous intensity; that is to say, clear places and places with shadows.

And that's the way, how the electronic eye of a camera, or a television camera can capture the different intensities in a person's face. And in order to avoid that the rays of light are reflected with the same intensity, it is necessary to apply a substance, that opaque the shining surface of the face of the person that is going to be shown before the camera. This is what is called makeup, because the areas of greater intensity are leveled with those of lesser intensity. But in short, the sum of these differences of intensities is what makes what we finally see is the disc of the Moon. And the surface of the Moon is an object that acts like a mirror, or has a surface, against which the rays of the electromagnetic waves that convert light are bouncing.

But if we go into smaller dimensions, for example the Moon being very tiny, this surface of the Moon will not be enough for a larger number of waves to bounce off it. In such a way that we will not be able to see the surface of the Moon. In this case we would have to place a detector, so that this detector can pick up the rays that we cannot see; and that it shows us for example, that the surface of the small moon is like a disc. But if we see shadows around it, such as when a lunar eclipse occurs, which has an effect like making up the face of the Moon, then we can say that the Moon has the shape of a sphere. But obviously if the Moon were very small, this detector has to be made by a substance, which in turn can capture those few rays that bounce with low energy against the surface of the imperceptible moon. But in this case, we can say that the Higgs boson was or is large enough for the detectors to "see" it, at the time of the rebound, when this particle caused a disturbance in the detectors. Or we might suspect that this detected particle does not actually correspond to the true Higgs boson. Because with the little amplified energy, it would be that it could be seen reflected before our eyes in a screen, or in a photographic plate, or another means, that made us deduce, that this was indeed a particle, and that by its low energy it corresponds to classify it as the boson of Higgs.

However, if the particles are very small, or let's say smaller than the photons of a ray of light, these rays will not be able to impact against those surfaces. So these rays so big with respect to some very small particles, will not be able to bounce, because they do not find a means or surface of support towards a detector, no matter how sensitive this one is. Therefore, we will not be able to see anything, because the energy is so tenuous that it is not enough to disturb the photomultiplier substance of the detector. In other words, these particles can pass through any detector, and will leave no trace for us to see indirectly their existence, and they will remain invisible. It is as if you were throwing a stone to try to hit the surface of a needle tip. And this stone is so big, that we won't have any information about what the center of the surface of the needle tip is like.

Or if we go to those very small dimensions, this is the reason why we have not been able to detect a large number of neutrinos, but in spite of their abundance, only a few have been captured by an immense tank of pure water that is located underground. For example, in the abandoned mines of Japan, where the Super Kamiokande laboratory is located. The Super Kamiokande observatory consists of an immense pond containing 50 million liters of pure water and is located one kilometer below the earth's surface. This pond is surrounded by some 11,000 photomultiplier tubes, arranged in a cylindrical structure, whose dimensions are 40 meters high by 40 meters wide. A muon is a massive particle. In such a way that rarely a muon interacts with water, and produces a well-defined signal. While electrons interact with pure water and produce as rains of additional particles. Therefore, the image detected by the 11,000 photomultiplier tubes will not be a definite signal, and the image we will see will be blurry.

But despite the very large dimensions of that detector, that will be a practical problem, so we conclude that we will not manufacture detectors to see the signal of almatrinos, because these particles are smaller than a neutrino. And if we have not been able to build detectors to capture neutrinos, then by the nature of the phenomenon, we are not going to be able to build detectors for almatrinos. Because almatrinos, although they are the most abundant in the Universe, are the smallest particles that exist, and for this very reason, that these are the particles that were initially formed, and that when they united gave the origin of the Universe. They form, for example, 74% of the undetectable energy of the Universe. But in addition, they came together to form an amount of matter that cannot be detected either, although this amount is as large as 22 % of the Universe. And to compare, we can only see 4% of that matter in the form of galaxies, stars and planets.

But going back to the errors of scientists, the movement itself of a particle, we owe it to the German physicist Ralph Kronig, who was the first to discover that particles have rotational movements, which is also called spin. But before exposing this in a conference, Ralph Kronig received a letter from Wolfgang Pauli, to explain to Kronig the need to assign to each electron of an atom, the four quantum numbers. This was one of the most important discoveries in physics, whose discovery we owe to the theoretical physicist of German origin, Max Karl Ernst Ludwig Planck, because it was Planck who discovered that the energy of electrons is quantized. In other words, only whole values can be assigned to this energy, which totally changed science's concept of energy and the structure of atoms. And the quantified energy could explain such transcendental facts as the ordering or location of atoms in a periodic table, and with this, we can deduce the behavior and combination of atoms in molecules to form matter. But also, that this quantification of the energy of electrons, was what marked the development of quantum physics, which represented another great advance in science, which was opening its way through a path that Max Planck pointed out to us.

In such a way that Kronig would have the idea that an electron, at the same time that it moves around the nucleus in its quantum orbit, can do it spinning around itself, just as the Earth does around the Sun with its movement of translation, and at the same time rotating in the form of spin. And that is why we have days and nights, whose duration is approximately 24 hours at the equator. Although in the poles a day, like a night can last six months, depending on the angle of inclination of the Earth. But perhaps because it is within the magnetic influence between Mercury and the Earth, and the great magnetic force of the immense planet Sun, Venus is spinning backwards. But the vortex shapes of galaxies tell us that they are turning counterclockwise, unless the photos are being looked at from behind. But Uranus has its equator rotated 90 degrees compared to the Earth's poles.

But Ralph Kronig would elaborate his mathematical model, in order to be able to explain the movement of the spin of a particle in itself. However, that this idea of Kronig, was something that would cause a great laugh to Wolfgang Pauli, since Pauli made known to Kronig, that this notion of rotation of an electron on itself, was without a doubt a ridiculous idea, reason why in the letter says Wolfgang Pauli to Kronig, and perhaps in a euphonic or burlesque way: "without a doubt that seems to me a very intelligent idea". Because Pauli also erroneously considered that with this mathematical model of the rotation of an electron on itself, he assumed that particles traveled at a faster speed than light, which violated Albert Einstein's law of relativity. And that, according to Wolfgang Pauli, was Kronig's mistake. And perhaps because he considered the great reputation of both Wolfgang Pauli and Albert Einstein, Kronig became discouraged, and made the great mistake of his life when he decided to take it back. So Kronig did not want to publish his ideas. But this was undoubtedly a great mistake of Ralph Kronig, because he was right; for a particle can, indeed, move faster than light.

But although it was also Wolfgang Pauli's mistake to refer to it as a Kronig atrocity, Pauli rectified, logically thinking, that Ralph Kronig was right. Because Pauli deduced that the movement of the electron should also have quantum values, which would lead him to deduce an idea; that by its logical nature, became a Principle. A principle that is rather based on a reasoned fact, but not on a mathematical model to describe it. Because it is to Pauli that we owe the deduction of the particle that we have identified as neutrino, but that discovery was not something theorized mathematically or by means of a theoretical model, but the sum of the energetic balance did not coincide.

And it was back in 1930 that Wolfgang Pauli, perhaps disconcerted because he did not find the solution to the phenomenon, proposed that there should be a particle in order to be able to compensate in the balance for the energy that was missing, so that the particle could not have charge or mass, since what was missing was only energy. And Pauli called this imaginary particle neutron. But this idea of a particle without charge or mass could not fit into Pauli's logic either, because it was difficult to imagine such a particle with such characteristics in those times. Until the Chinese physicist Wang Ganchang, proposes the idea of being able to detect this particle proposed by Pauli. And in 1956 the practical physicists Clyde Cowan and Frederick Reines, managed to elaborate an experiment to discover this particle. However, due to the fact that a particle called neutron already existed, the physicist Enrico Fermi, perhaps influenced by his Italian nationality, proposes to Wolfgang Pauli that he calls this particle rather neutrino, which means small neutron.

But returning to the case of the quantification of the electron in an orbit, Pauli definitively accepted Kronig's idea, and deduces that there have to be logical rules that describe the movement of an electron spinning in itself. And a series of imaginative restrictions are established, which are now called, as it was said, principles. And in this case this principle is known as the Pauli Exclusion Principle, which because of Kronig's error, or because it does not investigate further the nature of the phenomenon, is not called the Kronig Principle.

But the truth is that someone deduced that Kronig's theory was mathematically valid, or that it did not violate Albert Einstein's law of relativity, as long as the value of the quantum number was divided by 2. That is, 0/2, 1/2, 2/2, 3/2, 4/2, 5/2... And in this way, the concept of quantized energy was not violated, since the values 0, 1 and 2 are integers, while the others are fractions (+1/2, -1/2, +3/2,-3/2, +5/2,-5/2...). So 0/2=0 corresponds to the quantum value zero. While 1/2 is a fractionated value that can be positive (+1/2) or negative (-1/2), because the rotation of one of the particles, such as the example of the planet Venus, is influenced by the rotation of the other. And obviously for an electron we can consider only four possible associated values, which are the four quantum values who mentioned in your letter, Wolfgang Pauli to Ralph Kronig.

But it is because of this quality of the rotation of a particle on itself that light exists, because the photons that form light are bosons, so light can form and travel; or bounce off objects as separate rays, or in the form of beams of photons without melting with each other. That is why we can see objects, and for the same reason, there is all matter in the Universe, because fermions, when they rotate, create force fields that make some particles feel attracted to others. Let's say, that's why there are spirits, trees, insects, water, planets, air, atmospheres, stars, galaxies, and so on.

It means that if we imagine very small particles such as neutrino and almatrino, this phenomenon of a particle spinning on itself, is going to have an enormous importance, or that this movement will be transcendental for the formation of other kinds of energy, and in the behavior of the energy that is being transformed into matter, and everything that is being formed in the Universe. Or that the energy can travel in the form of polarized electromagnetic waves, that is to say, that an electric field and a magnetic field are formed in the same wave, because the electric and magnetic fields move at an angle of 90 degrees between them. In other words, without integrating as a single wave. And so the electromagnetic wave cannot pass through obstacles, in an electromagnetic phenomenon known as "faraday's cage".

And this is a fundamental condition for the formation of light. Because visible light is an electromagnetic wave that does not penetrate objects, but bounces against them, which is essential for the effect of the vision of the eyes, that is to say to be able to see objects, when the rays of light bounce and we can collect those rays through the retina. Or, as it was said, that television cameras and those that capture a photograph are based on the same principle. Or let's say, that this electromagnetic energy influences in an important way for life, and especially for the everyday life in human beings, as shown only a few cases in Figure 1.

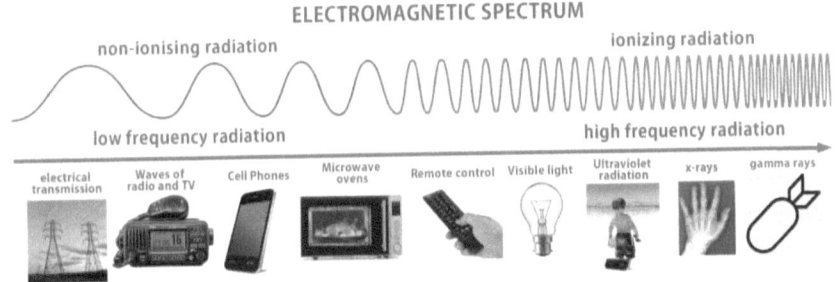

FIGURE 1
THE ELECTROMAGNETIC SPECTRUM WITH ITS WIDE RANGE OF ENERGY, AND ITS INFLUENCE ON THE BEHAVIOUR OF MATTER AND THE EXISTENCE OF LIFE

From here, Wolfgang Pauli deduces that integer values give different physical properties to particles with different quantum numbers; and to differentiate them from each other, one is called a fermion and the other bosons. And to the zero value of the bosons corresponds the particle of Peter Higgs. And so this particle became one of the most sought after, since this would imply that this was the particle from which God formed the Universe. Therefore, this particle would deserve the honor of being the particle of God, because it would be with which

God began the formation of the Universe. And Higgs concluded that it was from this particle that the Universe began to form.

But here is the other error of Peter Higgs, because what he never imagined, is that there are particles smaller than the boson of spin zero. And that the spin is but a form of rotation of a particle upon itself, and these particles can be as small as an almatrino, or as large as the Earth around the Sun, or the Sun itself spinning along with the Milky Way; and this galaxy is spinning from left to right around a cluster of suns. But the shape of the vortices of the galaxies indicate that the rotation of the galaxies is from left to right, or as does the Earth, as many claim it is from right to left. But whether it is one way or the other, this does not influence the idea we want to explain, because if two galaxies separate, one will turn to the left and the other to the right, as a natural consequence of the influence of the electromagnetic field.

4

THE MOMENT BEFORE THE UNIVERSE WAS FORMED

So, Wolfgang Pauli deduced, without having a mathematical model, that an electron having a quantum value of 1/2 can rotate indistinctly from left to right or from right to left as the Earth does. However, when there are two electrons at the same quantum level, the influence of the first can affect the second, because the electron rotation has created an electro-magnetic field, which makes this second electron rotate in the opposite direction to the first. That is to say from left to right or from right to left, for which, the rotation can take values (-1/2) or (+1/2). Because the plus and minus values are assigned relatively. While a boson cannot in itself create an electro-magnetic field, because the bosons do not have a specific sense in the rotation. So the first boson cannot affect a second boson that is in the same orbit. And these particles that have fractional values of their quantum values are called fermions in honor of Enrico Fermi.

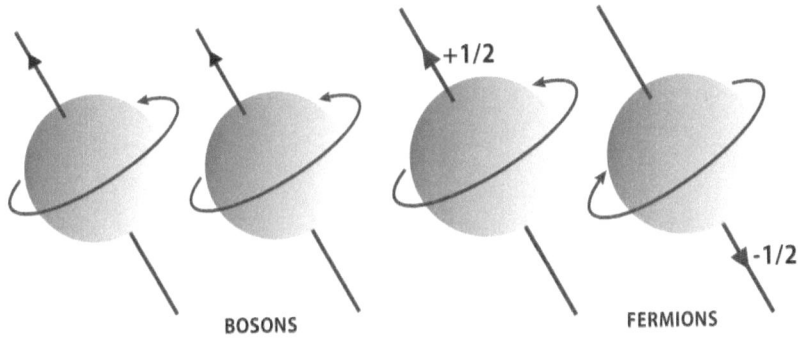

FIGURE 2
BOSONS CAN ROTATE WITH INTEGER QUANTUM VALUES, WHILE FERMIONS HAVE A FRACTIONAL QUANTUM NUMBER, AND THE ROTATION OF ONE OF THEM INFLUENCES THE DIRECTION OF ROTATION OF THE OTHER

But Pauli continues to deduce that if two electrons occupied the same orbit with the same quantum number, their form of rotation could not be in the same direction. So this would be an impossible movement, because a moving particle gener-ates, as said, an electromagnetic field, which will have an in-fluence on the other particle, as seen in Figure 3.

And it is from this rotating movement that the electrical charges of two poles are generated: the positive pole and the negative pole. In such a way that the Principle of Exclusion of Pauli, establishes in a logical way, that two fermions that turn in the same direction, cannot be occupying the same orbit, or that they have the same quantum number, because in a logical way that these two particles would condense, and would become a boson. But in any case, even hypothetically, this kind of movement in the same direction of rotation of two electrons in the same orbit would be an impossible fact. Or as the example referred, of Venus rotating in the opposite direction between the orbits of Mercury and the Earth.

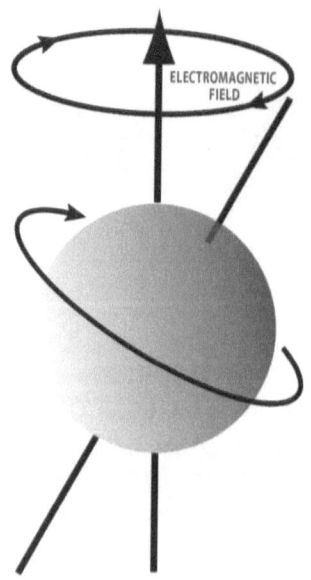

FIGURE 3
MOVING FERMIONS GENERATE WAVES
ELECTROMAGNETIC

The term boson was suggested by the English physicist Paul Adrien Maurice Dirac, when at Dhaka University in India, it commemorated the anniversary of the contribution of the

Professor of Physics from the University of Calcutta and Dhaka Satyendra Nath Bose. Bose was born to a middle-class Bengali family on 1 January 1894. And at an early age, Bose already showed signs of his genius. But an interesting anecdote is that this young student was given in his math grades a value of 110 out of the maxim of 100. And the ten extra points were given to him, because Bose not only answered questions correctly, but he answered other matters in more than one way. And in honor of Bose, it was the legendary physicist Paul Dirac who proposed the word boson for that particle that Bose discovered with his statistics, which was later called Bose-Einstein theory.

Like two integer particles, they do not generate electrical charges, so they may be occupying the same orbit. But they would merge forming another particle with greater energy; that is to say: 1+1=2 that would correspond to a boson of integer value 2. The value +1, -1, +2 and-2 although they can be considered mathematically, actually in a physical way that their direction of rotation would not have much importance, which is different from the particle rotating in the same orbit with values -1/2 and -1/2 that gives 1, or that +1/2 plus +1/2 is equally 1, which will be a boson because it corresponds to an integer value.

For example, the photons that form the light are bosons, which are also called particles of force, because they condense or integrate the energy, as for example the gluons, where apparently a third electric pole arises. Another example is the force of gravity; although to explain this the graviton is proposed. Or any nucleus that has as value of the spin an integer number, and obviously that the bosons do not fulfill the principle of exclusion of Pauli. And the other particle that also has spin zero, apart of the Higgs boson, is called pion. And the importance in life, whether it is called physical or spiritual life, is that electrons, neutrons and protons are fermions, while the photons that form a beam of light are bosons as mentioned, and bosons constitute the forces that integrate them. And in the nuclei are the gluons; that is, that the bosons prevent matter and light from disintegrating.

And depending on the intensity of these forces, the electrons remain rotating around the nuclei forming matter; that is to say, for example, all forms of life. In such a way, that this property of particles fermions and bosons, necessarily determine our form of life and in the life itself of the Universe, because the result of this interaction, is what forms a spectrum of a beam of photons to certain temperature of equilibrium, that possesses a spectrum of Planck. And an example of this is the radiation of the cosmic microwave background, which are the traces or witnesses that allow us to go back in time, to have an idea of how the Universe could have been at the beginning, or even before the Universe was formed.

It is by this natural principle that we say that the particles that created the universe cannot be bosons but fermions, because when the particles rotate, they create an electromagnetic field, as shown in Figure 3. So if the particles that formed at the beginning were bosons, matter would not have formed, because there would not be electromagnetic charges, also known as electrical charges that are the forces that maintain motion. For example, any electronic device, an automobile or the immense turbines of a hydroelectric plant, only work when an electronic charge flows through its circuit from the negative pole to the positive pole. Or as it was said, if it were not for the fermions, spirits would not exist. And in the Universe there would exist only one intelligence, which Peter Higgs would call God. But the truth is that there is matter, and infinities of beings that move as spirits or intelligent energies: call themselves geese, dogs, cats, fish, spiders, snakes, viruses, microbes, spermatozoids, plants, and so on. But there is also the Earth, so the Universe was formed from fermions but not only from bosons.

And some intelligences have become conscious of themselves, such as human beings. While others are learning to be, as in the case of monkeys, raccoons, badgers, sheep dogs, pigs, crows, elephants, cats, intelligent dolphins and moles. All of them show a degree of energetic mastery; and the ability to remember, which is basic, because memory is necessary for the evolutionary process. In such a way that almatrinos are actually fermions, because many forms of independent spirits were formed; and we conclude, that almatrinos cannot be bosons. And we can say that without bosons there would be no light; and without fermions there would be no matter, and without both particles there would be no Universe.

But the error of Peter Higgs, to which we referred, is that he did not consider the relationship of virtual numbers, because effectively, the problem will not cease to exist just because it cannot be represented by a mathematical function or formula. Because mathematics, as it was said, is only a tool that science uses to capture an explanation; and the solution of a phenomenon is in fact real or really real, and there is no other alternative.

So Peter Higgs' mistake was to consider $0*2=0$. But according to Higgs nothing less than 0 can exist, so, from there or from zero, the creation of the Universe should have happened. But also, that someone had to have intervened to begin this creation. But according to what we consider in the Book "The Universe before Time Zero"; 0*2=0 we can also write it as 0/0=2. But equally than 0/0=1 or 0/0=1/2, and this, of course, would not make any sense from the purely mathematical point of view, because it would be equivalent to say, that 2=1 or that 2=1/2 or ½=1. Or any division made by zero would give different values.

But the young Venezuelan mathematician Ramses Cornieles addressed this problem of division by zero, as we mentioned, and resolved this incongruence of division by zero. And Ramses represented for example, the value inside a circle, to indicate that this value is included inside a zero. In such a way, that now we can write the fractionated value of Peter Higgs as 0/2=Ⓞ. But this value cannot be zero, because it is contained within another zero.

And now we will not be able to say that the Universe began to form from the zero point, but from long before the zero, because we can include the value within another zero Ⓞ; and so on in a specular or virtual way. That is, we can write Ⓞ/2= ◎. A zero value included within another zero value, until we place ourselves in a more logical way in the range from the minus infinite to the plus infinite (-∞, +∞). And in the least infinite nothing existed and nobody from nothing could form the Universe, because there nothing and nobody could exist.

But perhaps, or rather, that this virtual analysis leads us at the beginning to what Paul Dirac theorized as an elementary particle of a single pole; that is, a magnetic monopole. Or a particle with "magnetic charge" in a magnetic field. Because what we have always known is the electric charge of an electric field.

Like, for example, the poles of a battery that give functional life to an electronic circuit, like saying a radio or TV, because the current or the battery have two poles.

But equally, we know that any magnet has two magnetic poles that we call north and south. But if we cut a magnet into two pieces, each part will still have its two poles, north and south. Or the same thing happens with chirality; because if you manage to draw a line through the mere center of your face, you will always have the left side on one side, and the right side on the other. But there has to be a line that does not have chirality, and that would be the mere center of your face, and in that line theoretically there is no chirality anymore. Therefore, or similarly to the poles of the magnets, as we cut the magnet more and more physically, there has to be a "magnet" that has only one pole, that is, north or south but not both poles. And this hypothetical substance will be a particle that will have a single magnetic pole, and Paul Dirac called it a monopole. That is why there is only one current of electrons in a copper wire, when there is the movement of bringing the wire closer to or farther away from a magnet. But the same happens if we move the magnet closer to the copper wire, or when we rub a silk cloth.

Although electric charges, they move better on the surface of noble metals, or those that have more fixed bosonic forces that integrate them. That is why the best metallic conductors are gold and silver, and electrons flow from the earth. Therefore, a large number of them can accumulate in gold or silver garments. And if people wear them on their necks as ornaments, they will become conductors of electric charges, so it is very likely that as a positively charged cloud passes over them, a current will be produced from the earth to the cloud, and the current will flow through the person's body, because the person with his or her gold edge bridges the electrons, and this person may die electrocuted, because for an instant too many electric charges passed through the conductors of his or her body. Mainly the heart cells that generate electricity with this movement and are the ones that keep the heart pulse active. A wet cow can also die electrocuted, because it was soaked in the hoofs that were the insulator, and the water conducts electricity, even though the cow does not wear a gold necklace. So it is also not good to place a metal bell on the cow's neck, because this will increase the risk that the cow will die electrocuted when the current of electrons flows from the earth to the cloud, using the cow's body as a conductor.

And these particles with a single magnetic pole exist, because we do not doubt that they are the almatrines, which formed the electromagnetic waves that spread throughout the Universe, and produce light, and a number of phenomena related to all existence. Because only one almatrino was necessary, which began to spin in the form of a spiral or faster and faster. And with this spiral-shaped spinning movement, the almatrino accelerated from zero, until it was fired with an enormous tangential force. And so the initial movement was created, as shown in Figure 4. And we imagine this view from above, because it is clearer to be able to draw a helix and see its effect tangentially. But this is a phenomenon observed in particle accelerators, where the force of rotation increases with the radius of the equipment. That's why the CERN particle accelerator has a circumference of 27 kilometers, and the Chinese are building an accelerator, whose circumference length will be 100 kilometers. And this accelerator may be completed by 2030. But unfortunately, although this accelerator is very large, there will be no detectors to pick up the signal that could come to us from the almatrinos.

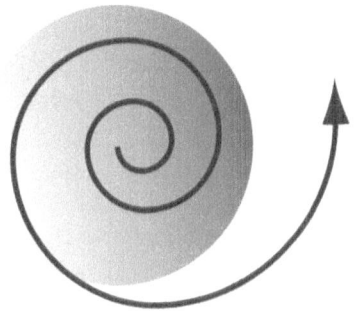

FIGURE 4
MAGNETIC MONOPOLE OF AN ALMATRINE SEEN FROM ABOVE

But let us say that the existence of magnetic monopolies was formulated by Paul Dirac in 1931, who did not accept the apparent irregularity shown by Maxwell's equations. However, by introducing into these equations the existence of magnetic monopolies, these equations would show a symmetry in the interaction between the electric field and the magnetic field, which would be what originated the electromagnetic field.

A magnetic monopole is a particle that has only one magnetic pole; that is, north or south but not north and south. And theoretically, of course, there can be a particle with a magnetic monopole, because the existence of this particle would be the basis to explain how the Universe originated from a single particle.

And on January 29, 2014, Professor David S. Hall of Amherst College Physics and the Academy Research Fellow Mikko Möttönen of Aalto University in Greater Helsinki, Finland, reported that they have managed to create, identify and photograph magnetic monopolies in the laboratory. And this, obviously, would give invaluable support to our theory of how the Universe was formed from nothing, because it only needed to form an almatrino with a single magnetic pole before time zero, as shown in Figure 4. In such a way that it becomes essential to write, formulate or extend a new form of the theory that reduces or eliminates doubts about the Big Bang.

And we can deduce, that the Universe began to form gradually or progressively from time zero, but that this was not a sudden arise, but that the Universe was gradually gestating. Where it would first begin as an embryo does, which is formed from a spermatozoon with an ovum inside the belly, but occupying a very minimal space. So in the same way, almatrinos without mass began to form, because this mass in repose of an almatrino m_0, now we cannot say that it is zero, but we can include it within a circle. Whose mathematical meaning is that this mass is not zero, because it is included within the value zero $(<m_0>)=m_0$. And this is a gradual progression, which complies with a natural principle, in which it is established that in Nature there are no sudden changes or jumps, but a continuity and contiguity of events.

And with the concept of almatrinos, virtual numbers and the magnetic monopole of Paul Dirac, from now on, we can already imagine how minimum space was before the Universe began to form; or what existed there before time zero. But everything indicates, that the Universe did not arise in an agitated way from a very hot point or high density, so it is necessary to apply the definition of this model to a new Big Bang, always in honor of Edwin Hubble, Georges Lemaitre and Paul Dirac.

And as for Albert Einstein's mistake, this happened, because he saw everything in a way relative to the phenomenon of light. It is even to Albert Einstein that we owe the explanation of the photoelectric phenomenon, whose principle is used in the amplification of the electronic current, in the more than 11 thousand photomultiplier tubes that are around the water pond of the observatory of the Super Kamiokande. But perhaps Albert Einstein's biggest mistake was to assume that nothing could travel faster than light, even though the experimental proof shown in Figure 5 indicates that light actually takes on infinite values, just like energy. And this great speed of the almatrinos in tangential form, is what forms and will form all the energy and mass that exists, and that which can exist in the whole Universe.

5

THE SPACE IN THE MINUS INFINITE

But the reality shown by the experiments is that a moving particle gains an additional amount of mass m from the resting mass m_0. So Albert Einstein made an important mistake, because he only relied on a mathematical function to state that if a particle moves faster than light, then the mass the particle acquires would be imaginary. Because mathematics indicated to Albert Einstein, that if the value inside the square root is negative, when extracting the square root, the quantity would be imaginary. But reasoning tells us that no movement can be imaginary. So Albert Einstein only deduced that mathematically, the mass m that the particle gains from its resting mass m_0 is given by the equation:

$$m = m_0 / \sqrt{1 - \text{U}^2/c^2}$$

In such a way that mathematically, U cannot be greater than C. On the other hand, Albert Einstein also assures that Energy cannot be imaginary, and the real phenomenon tells us that mass is formed from energy, so something that is derived from something real cannot be imaginary either. Only that Albert Einstein would not find a way to solve the negative value of this equation, and he considered from this function that nothing could move faster than light. And he only focused on seeing the problem in a mathematical way, but not on analyzing the phenomenon in a logical way. But this affected, as we saw, the idea of Ralph Kronig, who made his mistake by considering only the great prestige of Albert Einstein and Wolfgang Pauli, but not the phenomenon itself.

But also using the mathematical tools that led Rameses Cornieles to solve the problem of division by zero, we used the imaginary value "i". And before Ramses proposed it, we had already solved the problem of the imaginary value of the square root, and we adapted it to a condition that is more adjusted to the real phenomenon; reason why we arrived at the equation that we saw previously:

$$\text{U} = m_0 * C^3 / E$$

Being m_0, the mass of the particle at the moment of being without movement. And here is introduced the speed with which the particle moves; that is ʊ, when the particle is in motion, while C is a constant, which actually represents the speed with which the bosons that acquired mass move, i.e. a concentrated beam of photons in the form of visible light. But the value C, in this case would be a constant, so it is independent of ʊ. And ʊ depends only on the energy of the particle E and its mass at rest m_0. And light only manifests itself when the electromagnetic waves that were formed interact with the gaseous substances of the atmosphere of the planets, because these waves in the form of light are nothing but the waves derived from the electromagnetic waves that had previously been formed. And light was formed, from the energy in luminous form that is released, when the electrons of matter return to their fundamental quantum level, once they have been promoted to higher levels by electromagnetic radiations.

And as you can see in Figure 4, this high speed is independent of the speed of light, and occurred when a single particle was fired with a tangential speed. And it is from this speed that mass m began to be created, and then one after another, until the particles interacted, and a physical and electromagnetic condition was generated that continued to create mass and energy; until we arrived at the zero point of the Universe. But all this was independent of light, because the planets did not yet exist for electromagnetic waves to interact with atmospheres, and light could manifest. Of course, by that time, we didn't exist either. In such a way, that the Universe necessarily first had to go through a period of absolute darkness, until the largest and most solid bodies were formed from the energy that was transformed into matter.

But this equation $U = m_0 C^3/E$ or $E = m_0 C^3/U$ explains to us in a more logical way, how the Universe was formed, because the mass arose from a very small movement, and this generated an energy that was equally very small. Or smaller than zero, according to the deduction of virtual numbers. And with this new concept, we will no longer be able to say that the values were zero; because if we say, for example, that the mass is zero and not smaller than zero, this would make a real physical phenomenon disappear mathematically. And the equation that formed the Universe, now we can write it as:

$$E = m_0 \Psi / U$$

Where Ψ is the new constant that replaces the other constant C^3. And the equation that explains how the Universe was formed, we can write it in a more logical way as follows:

$$<(E)> \ = <(m_0)> * <(\Psi)> / <(U)>$$

And with this form, we can no longer say that in time zero the mass was zero, but that this mass did not exist, because it began to gestate from nothing. And maybe as it was said, everything started from a magnetic monopole, because it was only necessary that an just one particle with a minimum energy, entered in an accelerated movement. But this accelerated movement no longer stops, because it generates its own energy necessary to continue with its movement, which at the same time, managed to generate other particles. But such a small mass is only possible, because we can include it within a zero. In such a way, that a phenomenon that is real, we can no longer make it disappear mathematically.

But whether or not a particle travels at a higher speed than light is no longer a purely mathematical phenomenon, but depends on the dimensions of the particles we are considering, as well as the distance they have to travel; or at least that we use two variables so that we can compare our auditory and visual senses. For example, Galileo Galilei referred to large bodies, or the weight of spheres. Then Isaac Newton considered these movements and represented them in written form through his mathematical formulas. But with these formulas, he deduced, for example, the law of universal gravitational attraction, which was exactly what Galileo listened to in an auditory way, when he rolled the spheres along an inclined ramp. And to find out how the force of gravity originates, one looks for another boson by the name of graviton.

But then Albert Einstein went further, and studied the phenomenon of light, because it was what was visible and tangible to him. And it is for this reason that Albert Einstein, referring to Newton says to him: "forgive me Newton, but what you deduce, is not fulfilled for the photons that form the particles of light". Then Stephen Hawking arose and referred to the elementary particles, and said: "forgive me Einstein, but what you explain for the light, is not fulfilled for the elementary particles".

But another error of Hawking, is that he could not imagine, that there are particles smaller than the elementals, and that we have had to call them in another way, that is to say almatrinos. And those particles moved with such a high speed that at first it tended to have infinite value. So we can say that this is the absolute velocity of an elementary particle. And that in addition to creating the Universe, the almatrinos formed and will continue to form all the mass and energy of the Universe.

Or even light itself, because on entering into motion, these particles created what James Clerk Maxwell defined as electromagnetic radiations. But Paul Adrien Maurice Dirac considered that it was a Maxwell error, because he did not include in his equations the magnetic monopole.

But this was neither a philosophical nor a mathematical fact, because the formation of mass is a real fact, because it is what exists and it was what was experimentally demonstrated in 1914. Only that this phenomenon was forgotten, because it was Albert Einstein who buried it along with his error, that nothing could move faster than light. But with the new techniques, it could be demonstrated only by mathematical extrapolation, that particles do create mass, but that the relationship V/C also goes towards infinity, as seen in Figure 5; and that is what created the mass of the Universe.

And we say by extrapolation of the mathematical representation, because before the value V/C=0.5 of Figure 5, the function is a straight line with a low slope, which means that V=0.5C or that V=C/2. And this, what it means, is that a particle that moves with a speed of 150 kilometers per second, begins to create mass, but this mass is very small. Then, when the velocity reaches the value of 0.8, the relationship m/m_e makes $m=m_e*\infty$. Or that the mass gained by the particle quickly becomes large with respect to the dimensions of an elementary space. And so the mass of the Universe was created, from a cold point and particles that began to move faster than light. It is a phenomenon that we can now mathematically integrate into the range $(-\infty, +\infty)$.

But all this deduction is a consequence, or is based on mathematical data that could be captured in graphical form as shown in Figure 5 in 1914.

And it is really so, that besides explaining how the Universe was formed, we can also move forward in the timeline, using the example of the three fictitious characters of Galileo Galilei and the flash of the cannon, when we place the cannon of Galileo at a distance of 4.7 billion kilometers. And when we travel at that high speed, we will be able to see events in real time; but someone who is at zero speed, for example who is mounted on the Earth, will see those relative events so that, as if those events were happening in the future, but that the same are events of the past for someone outside the Earth to be able to travel faster than light.

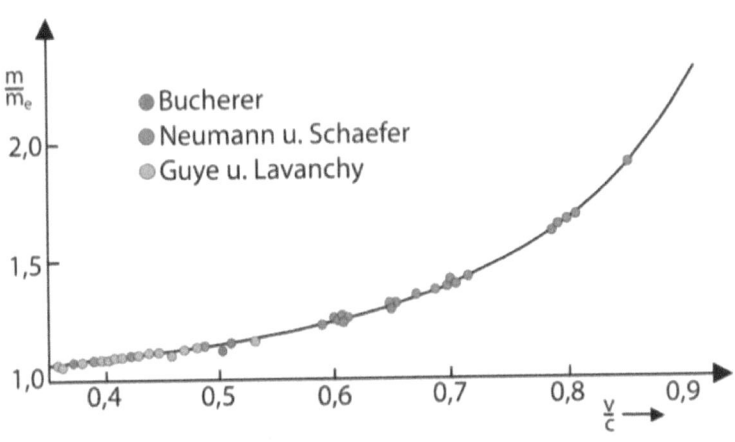

FIGURE 5
THE MASS THAT A PARTICLE GAINS WHEN IT IS ON THE MOVE

Or take a baseball catchers, for example, which is receiving the pitches that a pitcher is sending out with a ball that goes well into the catchers at 150 kilometers per hour; that is, at about 40 meters per second. The catchers will be able to see that the throwing event happens very fast, while we, if we manage to ride on the ball, will notice that the time has not elapsed, because our speed is zero with respect to the movement of the ball. Although we are moving along with the ball at 150 kilometers per hour.

And now we can say that the maximum speed at which a particle can move is actually an absolute value, which is mathematically equal to 27,000,000,000,000 kilometers per second. And when a particle moves with this speed, it is going to be totally difficult to detect it. But we will have to look for new mathematical models that can describe or embody the description of these cosmological phenomena.

Because even if we disintegrate a magnet or the mass m_0 as many times as we can think of, it can never be zero in the equation, but it will always be smaller than the zero of zeros. Or you won't be able to locate the line that you mark on your face, where it starts and where the left-handed part and the right-hand part end. And in this way, the mass m_0 can continue appearing in a successive way, as the mass m_0 within the mass of another zero ($\textcircled{0}/0=0$), and so interminably or indefinitely towards the value (if it can be called that way) minus infinite ($-\infty$), because the limit of the smallest can now be imagined by us in both physical and mathematically modeled form. The same situation occurs with space, which will now be the smallest place that can fit in our mind. And that ability to imagine the smallest things, is what makes us think, that we really come from a Micro World.

But in practice it does not mean that the physical phenomenon does not exist, or that it has to disappear obligatorily, because mathematically its nature cannot be explained. But what is really true is that the energy and mass of the Universe exist, and will continue to exist, as long as the Universe remains in motion. But something as immense as the Universe, nothing will be able to stop it, and we will not be able to do absolutely anything to stop this movement. In such a way, that only what we have left is to be able to live rejoicing that we belong to the Universe, and that all living beings have the same right to live in the Universe, but this is not an exclusivity of human beings, when they believe that someone granted them that right, and that for example animals and plants do not have those same privileges.

6

EXTENDING THE BIG BANG THEORY

It is through the concept of virtual numbers that we can imagine what size space was before zero time, since there something too small began to form, to be able to assign some dimensions to it, or it would be the equivalent of dimension zero. But then the system began to move, until it reached time zero; that is, the moment at which enough interactions were made. And when that tiny system reached that zero point, it is from here that we can begin to count the time of a new Big Bang, which we can extend to a time before zero. Because it was in that instant, that the critical energetic force was produced, that made the small system no longer support the high energies generated relative to the small space, because these forces were accumulating before the formation of the Universe happened.

We say that these were high energies, because they corresponded to the size of that point, but although the heat was infinite with respect to that tiny point, if it happened for example in the tip of our index finger, surely we would not notice that there was something hot there. But this is how the conditions were created for the Universe to begin to form from that place in the initial time. And this happened gradually but not suddenly or spontaneously from time zero; or it is from there that we can begin to count the time zero of the Big Bang. And obviously, that the great majority of scientists only want to explain the phenomena by means of a mathematical formula, and in this case, the relation between the mass and its volume is the density, that is, $V=m/\rho$.

And the only way to explain where this mass originated is to erroneously assume that the density ρ was very large at that time, because in that very small volume was concentrated all the mass of the Universe. Because it is thanks to Albert Einstein's mistake that scientists made another mistake, when they didn't notice, that the mass m is actually formed from the movement of particles. But before that critical moment, in the time zero, in reality the particles did not have mass, and a single particle of a single magnetic pole was formed that we have had to call almatrino, because the space to lodge that particle did not have volume; reason why, the density could not exist either.

And as for the high temperature, well, we already explained that the Big Bang as it is, also does not explain where the heat that heated this point in a time that is not zero but 1×10^{-35} seconds, which is the minimum value that can be assigned as the Planck time. But at the same time, we would have to formulate another concept of time, to describe the lapse between the minus infinite range to zero point, that is $(-\infty, 0)$. Although this concept of time should rather be defined as an eternal moment, because it does not change and still exists, as long as the Universe exists.

But although this model can effectively offer an explanation, as is the abundance of the elements, the Big Bang left us a trace, as is the cosmic microwave background, and also the law that Edwin Hubble discovered. But if these observed conditions were extrapolated back in time, that is, using only the known laws of physics, the prediction would tell us that just before a period of very high density and high temperature, in that way we are not going to be able to explain or understand, with this same model, how these conditions were really arrived at. And the discrepancy of this sequence of events and forecasts, has been catalogued as "one of the worst predictions that has happened in the whole history of physics".

Thus, when the Universe was believed to be static, this happened for a long time, because there was no formula to describe this event in any other way. It was similar to riding imaginary mounted on the ball thrown by the baseball pitcher, where we have the feeling that the ball is motionless, even though we move with it with a speed of 40 meters for every second passed. And this was thought to be so, until Edwin Powell Hubble, was able to look outside the ball, and located a point of reference, and realized that the galaxies are moving away from us, who are ridden on Earth.

And so Hubble observed that the lines of the electromagnetic spectrum that we see in Figure 1, is toward the red, in that narrow range that corresponds to the visible part of that immense spectrum. Because Edwin Hubble deduced that if galaxies were approaching us, such a shift would be towards a visible zone but the one that corresponds to the color blue. But in reality we are not going to be able to see absolutely anything either below or above this narrow visible range.

To see or pick up electromagnetic waves below or above that range visible to the retina of the human eye, we would have to use the right equipment: for example, a device that picks up very low frequency waves, such as a receiver that intercepts the waves sent by a radio wave source; or a television device that sees the images that we will not be able to see. Or a device that we use as WIFI, dark lenses to attenuate ultraviolet radiation from the Sun, etc.. But we could not be very close, when the explosion of an atomic bomb occurs, because these radiations have so much energy that they can pass through the cells, and can damage the DNA, because this range is very high energy corresponding to ionizing radiation of gamma waves. However, what we will not be able to say is that there are no radiations with a greater energy than the gamma, because we still do not have a system to be able to detect those radiations. Because in reality these waves have too high an energy, and they would be similar to the waves or radiations that were formed at the beginning in the Universe.

But ultimately, it was thanks to that keen observation of Hubble that opened the imaginative mind of scientists. And perhaps the one who was most interested in it, as mentioned, was a religious, the priest Georges Lemaitre, who pointed out, based on Edwin Hubble's observation, that if the Universe really is in full growth, there would necessarily have to be a point, from which all this event of the growth of the Universe originated.

Until 1964, the footprint, i.e. cosmic microwave background radiation, was discovered, which was undoubted evidence predicted by the hot Big Bang model. Since this theory, considers the existence of background radiation throughout the Universe, long before such radiation was discovered. The problem is how was it said, how, or from where did this heat originate? Or also that the discovery of cosmic acceleration in 1998, continues with the interest of finding somehow the cosmological constant.

But we only hope that with our theory of the gestation of the Universe, the agitated raft loaded with scientists, philosophers and religious, will definitely enter a sea of calm, so that humanity will give more value to its existence, and to the existence of all the other beings that inhabit the Earth. Because absolutely, we all have the same right to live, because absolutely, we all arise from the same energy that formed the Universe; that is, we should all enjoy the way of life that corresponds to us, but without the need to be harassed or harass each other, or that we continue killing our brothers and sisters animals to feed ourselves with the flesh of their bodies, because that really is not necessary, and is contrary to any Law of natural origin.

But perhaps one day, and with the torch of this knowledge, we will be able to illuminate the darkness in which humanity is enclosed, so that this incarnated form of life will come out of its tenebrous phase, in the same way that the Universe came out of darkness, when light began to form. And may this only be a phase through which humanity would have to pass, so that, as Stephen Hawking mentioned, humanity may carry the torch of knowledge to the highest level, and thus enter into a new stage of consciousness, which is a necessary matter for its existence, and for the existence of all living beings.

However, going back to the analysis of the Big Bang, this theory was not complaisant to the great cosmologists, which was undoubtedly another big mistake, because many of them reasoned, that by the fact of having begun or having an origin, instead of being stationary, the Big Bang theory was supposed to incorporate those religious aspects into science. Because someone was supposed to intervene to start that growth. But perhaps, that was only a coincidence, since that was the reality that fed more the doubts of the cosmologists who are still rowing in the same raft. Since they in the turbulence, try to separate the scientific thought of a real phenomenon, but that logically goes by the two slopes of the philosophy, as it is the science and the religion. And one feeds back on experimental proof, while the other is based only on a philosophical idea. And the philosophical idea will have to disappear along with its doctrine of philosophy. But the problem is that all this is part of human thought when it tries to investigate in order to elaborate an explanation. So we will not be able to separate those forms of thought, only by the fact that someone took for the explanation of the same phenomenon, a different way. Because Hubble, for example, was an outstanding sportsman, and apparently his father was religious and wanted his son

Edwin to be a reverend as well. While it is proven that the creator of the Big Bang theory, Georges Lemaitre was a Catholic priest. And cosmologists are only cosmologists, but what we are not going to be able to doubt, is that we are all sailing inside the same raft.

And we can nevertheless change the philosophy of thought, but the origin of the phenomenon and its logic is the only thing that we are not going to be able to change, and it doesn't matter if we are scientific or religious. But the example is, that despite being a religious, Lemaitre thought, in a well-reasoned way that:

"If the world has begun with a single quantum, then the notions of space and time would have no reason at first; and they will only begin to have a meaning when the original quantum would have been divided into a sufficient number of quanta. And if this suggestion is correct, the beginning of the world happened a little before the beginning of space and time".

But this surprising appreciation of Georges Lemaitre is correct, but it was undoubtedly what led us to explain what the Universe was like before time zero. Only that the Universe, as we have shown, could not start at the zero moment, at a point with a high density, but also extremely hot, because this presupposes the existence of an energy prior to the event. Therefore, that would not explain the existence of energy and dark mass, which is one of the mistakes that has to be faced if we follow the Big Bang theory. And our theory of how it began to form in the universe from nothing acquires more force. But if it happened any other way, let everyone mention their logic, because Ptolemy's logic was alive for more than 1500 years.

Although it all forms part of the reasoning capacity of the human being, regardless of the line he has chosen as his labor to achieve his own reasoning. But what we need from now on, since the growth of the Universe is just beginning, is that a change in the consciousness of the human being is necessary or has to occur. Or if the gestation of the Universe took 0.75 billion years, that is to say 9 cosmic months, the Universe is only an adolescent of 13.8 billion years. Which means that we would still have a long way to go to learn to live without making the same mistakes of existence. But it is necessary and urgent to raise the consciousness of the human being, so that humanity can correct its behavior in time, before humanity inevitably destroys itself. For if the raft were made of wood, humanity is devouring its own raft as if it were a swarm of termites.

And with the emergence of a new amount of heat Q, it will become greater and greater, but this enormous amount of heat generated will be transformed into a greater amount of mass, according to the equation that defined the theory of relativity: $m = m_0 + Q/C^2$, or $Q = \Delta m C^2$. Or in the same measure, or whenever a new amount of energy is formed, according to the Russian professor Andrei Linde, whenever a new amount of heat appears, a new amount of mass will be formed, and new galaxies will appear; and this is the only way, that the amount of heat generated, is appeased from that enormous amount of energy, when it is condensing into the form of mass. Because the condensed mass can store an enormous amount of energy. Or let's take the example of an atomic bomb, or gasoline that is nothing more than liquid energy that we can carry in the tank of our vehicle, to cover a great distance, and so on.

Therefore, with our analysis of the almatrines, we can now understand why the growth of the Universe happens in an accelerated way. But it also explains Hubble's other observation, with which he realized that galaxies actually formed from clouds in the form of cosmic dust.

And with a new theory adapted from the Big Bang, that can now offer us a broader explanation of a range of observed phenomena, including the abundance of light elements such as hydrogen and helium or lithium, and perhaps most importantly, the Big Bang theory is based on Albert Einstein's model of the theory of general relativity. But it will help us clear the way for other assumed theories, such as the homogeneity and isotropy of space or the deformation of space-time, because time does not exist. But the mathematical equations that explain or stamp these observations were formulated by the Russian-born physicist and mathematician Alexander Friedmann, so another must appear as Friedmann who mathematically formulates the new theories.

And between 1968 and 1970, Roger Penrose, Stephen Hawking and George F. R. Ellis, published works in which they demonstrated that mathematical singularities were an inevitable initial condition of the general relativistic models of the Big Bang. And then, from the 1970s to the 1990s, cosmologists worked on characterizing the Big Bang universe and solving outstanding problems.

In 1981, Alan Guth made another breakthrough in theoretical work on the resolution of certain problems related to the Big Bang theory by introducing a time of rapid expansion into the early universe, which he called "inflation". Meanwhile, that during these decades, there are two questions in the formulation of cosmology that generated discussion and disagreement, such as that on the precise values of the Hubble Constant and the density of matter in the Universe, before the discovery of dark energy, which was considered as a key prediction for the final destiny of the Universe.

And since the late 1990s, other significant paths in Big Bang cosmology have been cleared, as a result of advances in new telescope technology, as well as accurate analysis of data from observation satellites. And cosmologists now have fairly reliable and accurate measurements of parameters to analyze the Big Bang model.

Yet despite this, in November 2019, Jim Peebles, Nobel Prize in Physics 2019 for his theoretical discoveries in physical cosmology, in his presentation of awards, pointed out that he did not support the Big Bang theory, due to the lack of concrete evidence of supports, and therefore Peebles stated that:

"...it is very unfortunate that one thinks of a beginning, whereas in fact, we do not have a good theory of something like the beginning".

But this is a mistake by Jim Peebles, because we have already demonstrated what that principle was like, and the only thing that would be missing would be for the theoretical physicists of the new generations to devote themselves to translating all that has been said into a single equation to expand the new Big Bang. Because perhaps, in order to explain this, we have to resort to a new mathematical model, in which the reality of the phenomenon, of how the Universe was formed, is adjusted in a more precise way. Because it is important to know how earthlings, where we come from and where we are going, to see if we can learn to live as human beings; that is, without wars between brothers and between human beings; but also to know how to value in the same way our brothers the animals, or our other brothers who are alive but cannot walk, such as trees, because they are necessary to form the forest that gives them shade, and under which other animals live, but in addition, the trees are fed with water that is not contaminated. But it is the trees, those who clothe the Earth, with the greenery of the most beautiful garment that can exist in this immense Universe.

ABOUT THE AUTHOR

Graduated from the School of Chemistry, Faculty of Sciences of the Central University of Venezuela, with a degree in Chemical Technology. Postgraduate studies in Food Science and Technology. Special work on the chemistry of natural products and the chemistry of diseases. Designer of chemical processes. Books: "The Chemistry of Cancer". "The Chemistry of Diabetes". "The heart attack". "Alzheimer's disease". "The Chemistry of Arthritis". "The Chemistry of Thought. "The Chemistry of the Spirit". "How the Universe was formed. "The expensalists". "Why you shouldn't eat meat. "The Micro World. "Does God Really Exist? "Objecting to Albert Einstein's Relativity. "Guessing the Future", "The Universe Before Time Zero"...

www.ingramcontent.com/pod-product-compliance
Lightning Source LLC
Chambersburg PA
CBHW030941240526
45463CB00015B/907